DATE DUE			
Feb 27			
Mar 13			
MAY 1			
JAN 15			
1-29-98			
2-12-98			
2-26-98			
4-16-98			
OCT 26 1998			
NOV 23 1998			

970023

E
560.9
HIG

Higginson, Mel.

Scientists who study fossils

Harding Academy
Elementary Library

#319

Memphis City School
Title VI 97/170

CORDOVA-K

596373 01075 50378B 00721E   003

SCIENTISTS WHO STUDY
FOSSILS

Mel Higginson

The Rourke Corporation, Inc.
Vero Beach, Florida 32964

© 1994 The Rourke Corporation, Inc.

All rights reserved. No part of this book may be reproduced or utilized in any form or by any means, electronic or mechanical including photocopying, recording or by any information storage and retrieval system without permission in writing from the publisher.

Edited by Sandra A. Robinson

PHOTO CREDITS
© Mel Higginson: cover, pages 8, 10, 12, 13; courtesy of South Dakota Tourism: title page, page 21; courtesy of The Mammoth Site: page 18; courtesy of Paul Sereno and the University of Chicago: page 4; © The Royal Tyrrell Museum of Palaeontology: pages 7, 15, 17

ACKNOWLEDGMENTS
The author thanks the Field Museum, Chicago, Illinois, for its cooperation in the preparation of this book

Library of Congress Cataloging-in-Publication Data

Higginson, Mel, 1942-
   Scientists who study fossils / by Mel Higginson.
     p. cm. — (Scientists)
   Includes index.
   ISBN 0-86593-375-8
     1. Paleontologists—Juvenile literature. 2. Paleontology—Vocational guidance—Juvenile literature. [1. Paleontologists.
2. Paleontology—Vocational guidance. 3. Occupations.
4. Vocational guidance.] I. Title. II. Series: Higginson, Mel, 1942- Scientists.
QE714.7.H54   1994
560.9—dc20   94-6995
                                                 CIP
                                                 AC

Printed in the USA

TABLE OF CONTENTS

Scientists Who Study Fossils	5
What Fossil Scientists Do	6
Kinds of Fossil Scientists	9
Where Fossil Scientists Work	11
The Importance of Fossil Scientists	14
Studying Ancient Animals	16
Discoveries of Fossil Scientists	19
Learning to Be a Fossil Scientist	20
Careers for Fossil Scientists	22
Glossary	23
Index	24

SCIENTISTS WHO STUDY FOSSILS

The last dinosaurs died millions of years ago. That leaves scientists with a problem if they want to study dinosaurs.

Scientists would love to have walking, talking dinosaurs to study. However, they have the next best thing — dinosaur remains, called **fossils.** Most fossils are either the bones, teeth or print marks of ancient animals. The remains of some ancient plants are also preserved as fossils.

The scientists who study dinosaur remains and other fossils are called **paleontologists.**

Fossil scientist Dr. Paul Sereno, a dinosaur expert, works at a dig for dinosaur fossils in Niger, Africa

WHAT FOSSIL SCIENTISTS DO

Fossil scientists know a great deal about the plant and animal life of the distant past. To continue learning, they collect and examine fossils.

Sometimes scientists find fossils lying on the ground like fallen apples. More often though, they have to use tools to remove fossils from soil or rock.

Scientists study a fossil to find out what kind of plant or animal it was. They also date the fossil, finding out its age. Fossils are often millions of years old.

A fossil scientist hammers at rock to free a fossil

KINDS OF FOSSIL SCIENTISTS

Fossil scientists have special interests and jobs. A fossil scientist may be an expert in the ancient life of a place, such as the LaBrea Tar Pits in Los Angeles, California. Thousands of ancient animals were trapped and preserved in LaBrea's tar pools.

Other fossil scientists may study a certain kind of dinosaur or group of dinosaurs.

Fossil scientists work closely with geologists, scientists who study rocks and soil. They also work closely with scientists who study living plants and animals.

A fossil scientist inspects long, saberlike teeth of saber-toothed cats that died in LaBrea tar 10,000 to 40,000 years ago

WHERE FOSSIL SCIENTISTS WORK

Fossils are found in nearly every part of the world. Fossil scientists visit nearly every part of the world, too.

Fossil scientists like to be "out and about." Some of their work, however, is indoors. Scientists work in **laboratories** to prepare fossils for collections. They also work in their offices and study in libraries.

Dr. John Bolt of Chicago's Field Museum examines the bones of an amphibian that died perhaps 340 million years ago!

Sharp eyes, hands, a pick and a magnifying instrument help a fossil scientist separate fossil bone from rock

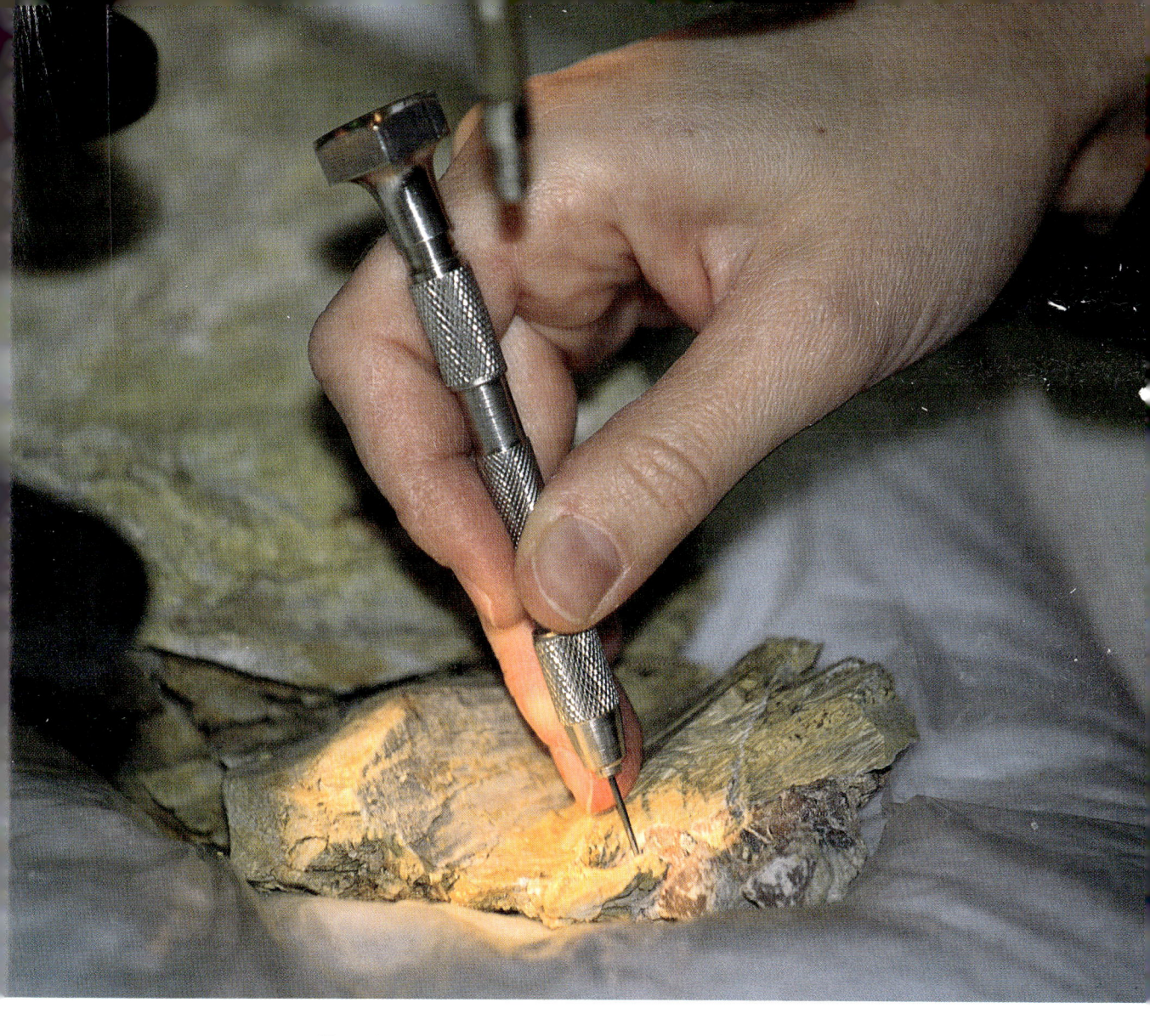

Fossil scientists, like dentists, must use their eyes and hands together with great skill

THE IMPORTANCE OF FOSSIL SCIENTISTS

Fossils are small — but important — clues to solving the giant mystery of life in the past. The fossil scientist's job is to help solve the mystery.

By studying fossils, scientists try to figure out what life in the distant past was like.

Each discovery by a fossil scientist is important. It can help everyone learn more about ancient life on Earth.

This painting at the Royal Tyrrell Museum of Palaeontology in Alberta, Canada, shows an Albertosaur and how life on Earth may have looked 100 million years ago

STUDYING ANCIENT ANIMALS

Dinosaurs weren't the only ancient animals, of course. Many other animals that seem strange to us today became **extinct** like the dinosaurs. Some, like the saber-toothed cats and woolly mammoths, disappeared "just" 10,000 years ago.

Dinosaurs are still the best-known ancient animals. New discoveries of dinosaur fossils may give clues to an unsolved mystery: Why did the dinosaurs vanish?

Lifelike models of ancient animals, like this Ankylosaur, can be made after scientists study the animals' fossil remains

DISCOVERIES OF FOSSIL SCIENTISTS

Fossil scientists have used their discoveries to tell an amazing story. Their story is about how, when, and where ancient plants and animals lived. Each "chapter" in the scientists' story covers millions of years.

One of the first chapters tells about dinosaurs and reptiles with wings. A later chapter tells about camels, sloths, elephants and dog-sized horses that roamed North America.

The story isn't finished. Each discovery adds something new.

The skeleton of a woolly mammoth, an ancient elephant, rests in rock at South Dakota's Mammoth Site

LEARNING TO BE A FOSSIL SCIENTIST

Fossil scientists usually have at least four years of college.

They study the Earth and learn about places where fossils are likely to be.

Fossil scientists study living plants and animals and learn about their **structure.** Knowing how modern plants and animals are "built" helps scientists recognize the ancient **ancestors** of those plants and animals.

Fossil scientists and their students hunt for fossils in the Badlands region of South Dakota

CAREERS FOR FOSSIL SCIENTISTS

Most fossil scientists work at universities. They collect and study fossils, and they teach science classes.

Many other fossil scientists work for museums. People working with the scientists help them prepare, show and preserve fossils. Museum scientists are often in charge of wonderful fossil collections. Some of the finest are at the American Museum of Natural History (New York), the Los Angeles County Museum (California), the Field Museum (Chicago, Illinois), and the Royal Tyrrell Museum (Alberta, Canada).

Glossary

ancestors (AN ses terz) — ancient relatives of people, plants or animals; the earliest people, plants or animals

extinct (ex TINKT) — no longer existing

fossil (FAH suhl) — the ancient remains of plants and animals

laboratory (LAB rah tor ee) — a place where scientists can experiment and test their ideas

paleontologist (pay lee en TAHL uh gist) — a scientist who studies fossils

structure (STRUKT chur) — the skeleton or basic "building blocks" of a living thing

INDEX

ancient life   14
animals   5, 6, 9, 16, 19, 20
bones   5
camels   19
cat, saber-toothed   16
college   20
dinosaurs   5, 9, 16, 19
elephants   19
Field Museum   22
fossil collections   22
fossils   5, 6, 11, 14, 16, 20, 22
fossil scientists   6, 9, 11, 14, 19, 20, 22
horses   19
laboratories   11
LaBrea Tar Pits   9
libraries   11
Los Angeles County Museum   22
mammoth, woolly   16
museums   22
North America   19
paleontologists   5
plants   5, 6, 9, 19, 20
print marks   5
reptiles   19
rock   6, 9
Royal Tyrrell Museum   22
sloths   19
soil   6, 9
tar pool   9
teeth   5
tools   6
universities   22